I0478459

# Our Insect Helpers: Hunters and Pollinators

## A Coloring Book

# Our Insect Helpers: Hunters and Pollinators

## A Coloring Book

By Michael Reed

Copyright ©2017 by MR

All rights reserved

## Acknowledgment

I  thank God for the interests and information for this book

Although there are   critters such as mosquitos, locusts, cockroaches, and bed bugs that are   harmful for our lives, some insects are useful to us by helping to control pests and produce our fruits and vegetables.

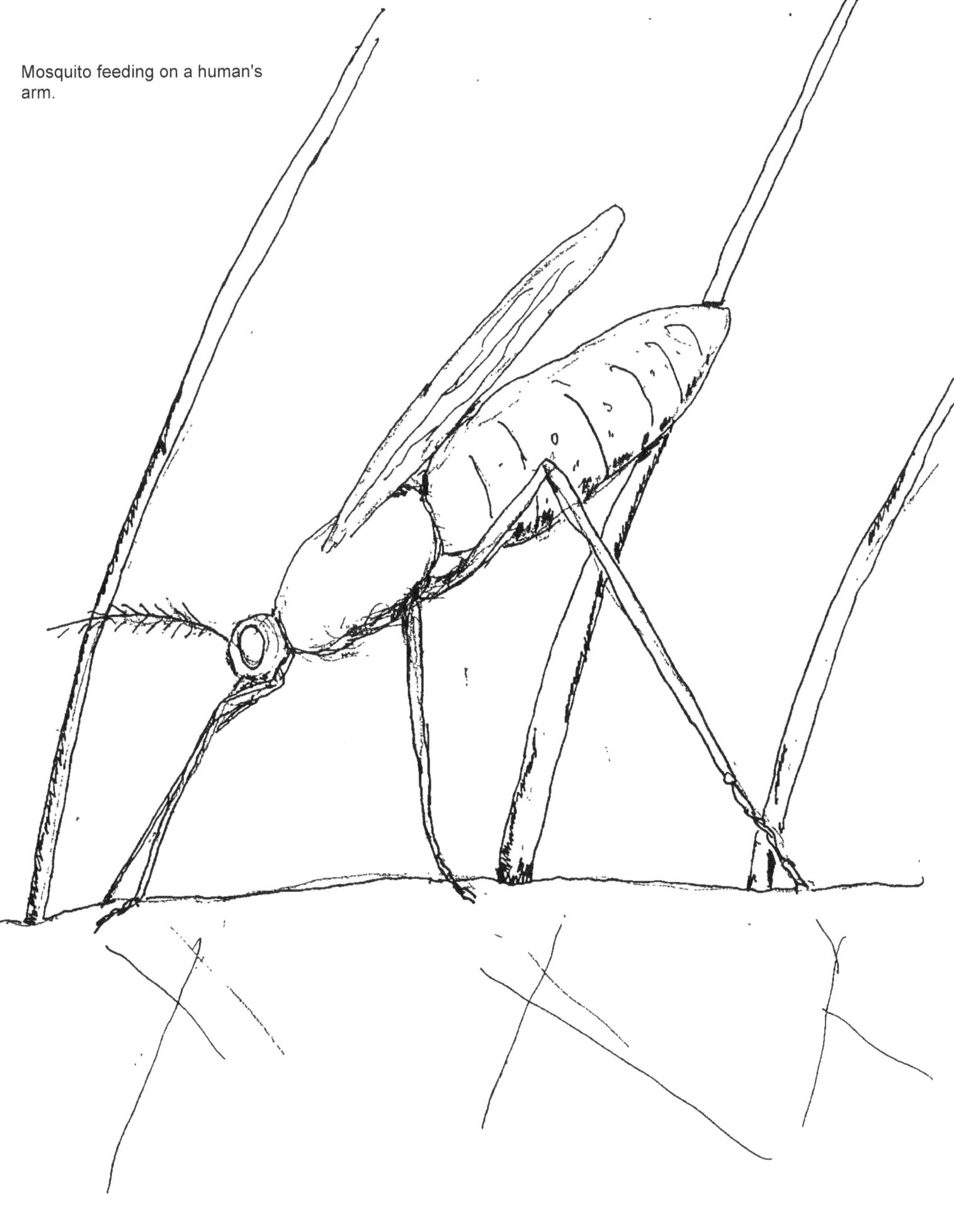

Mosquito feeding on a human's arm.

Cockroach

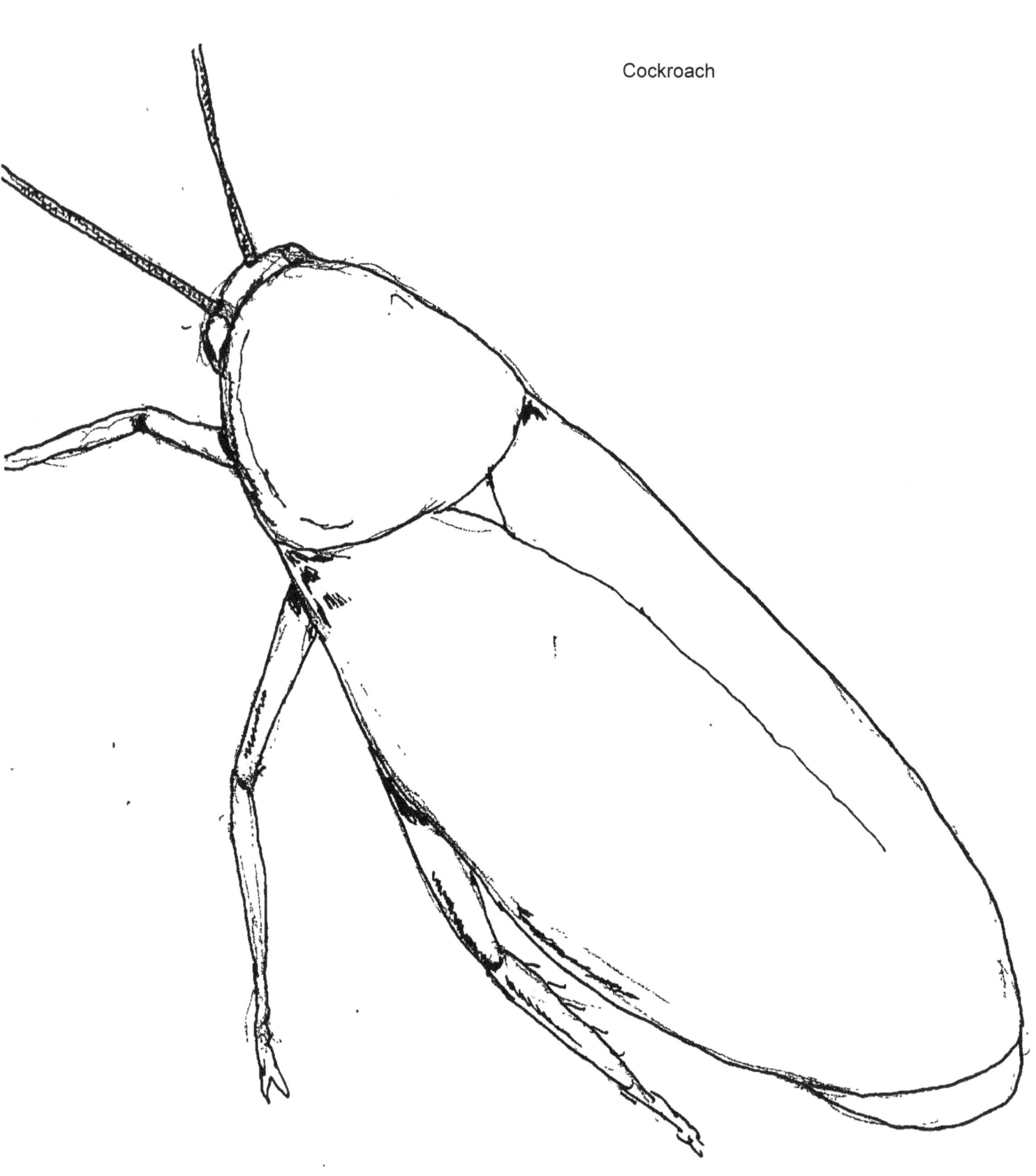

Two bed bugs on a human's skin.

A swarm of locusts on a field.

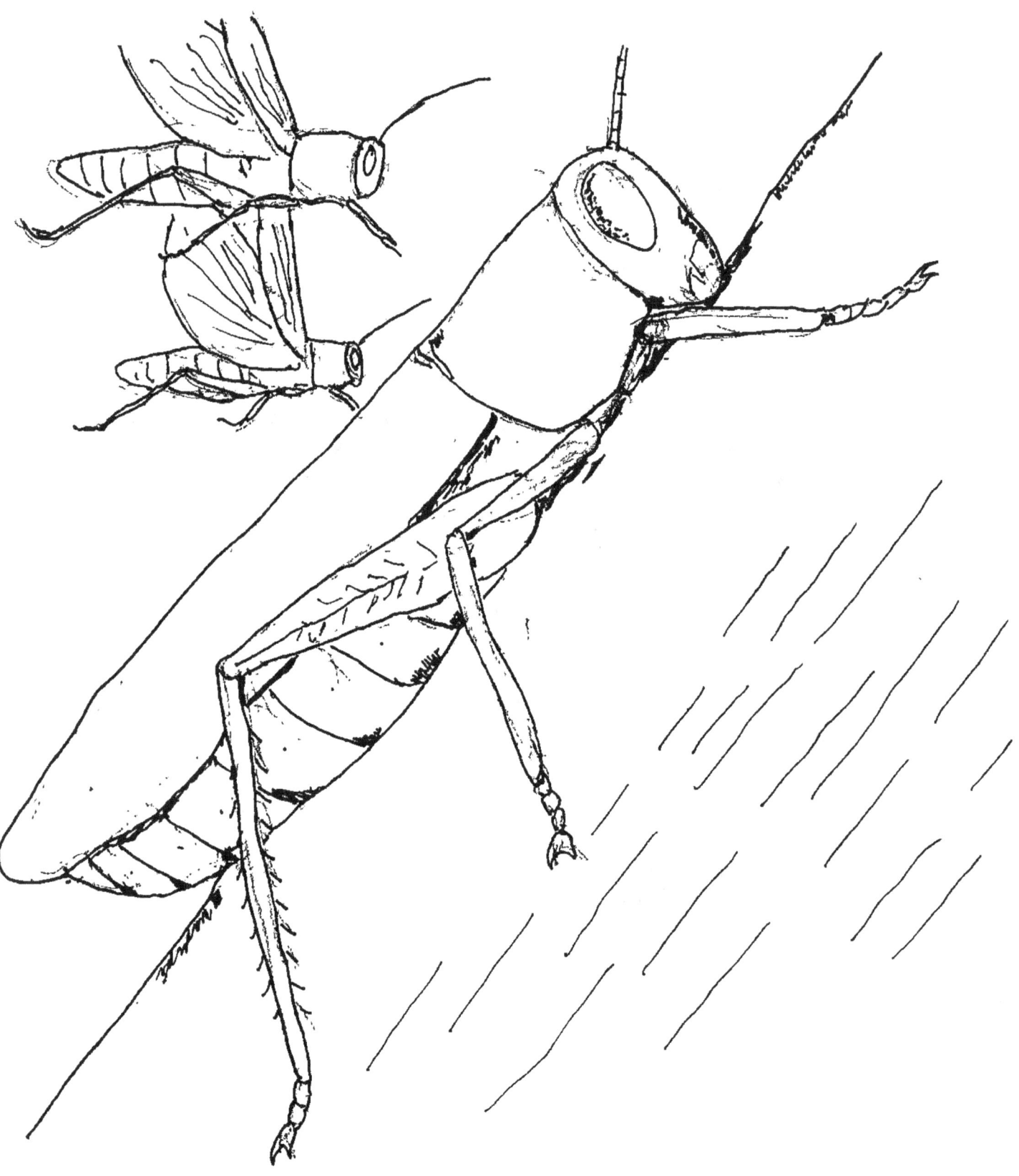

A praying mantis having a fly for lunch.

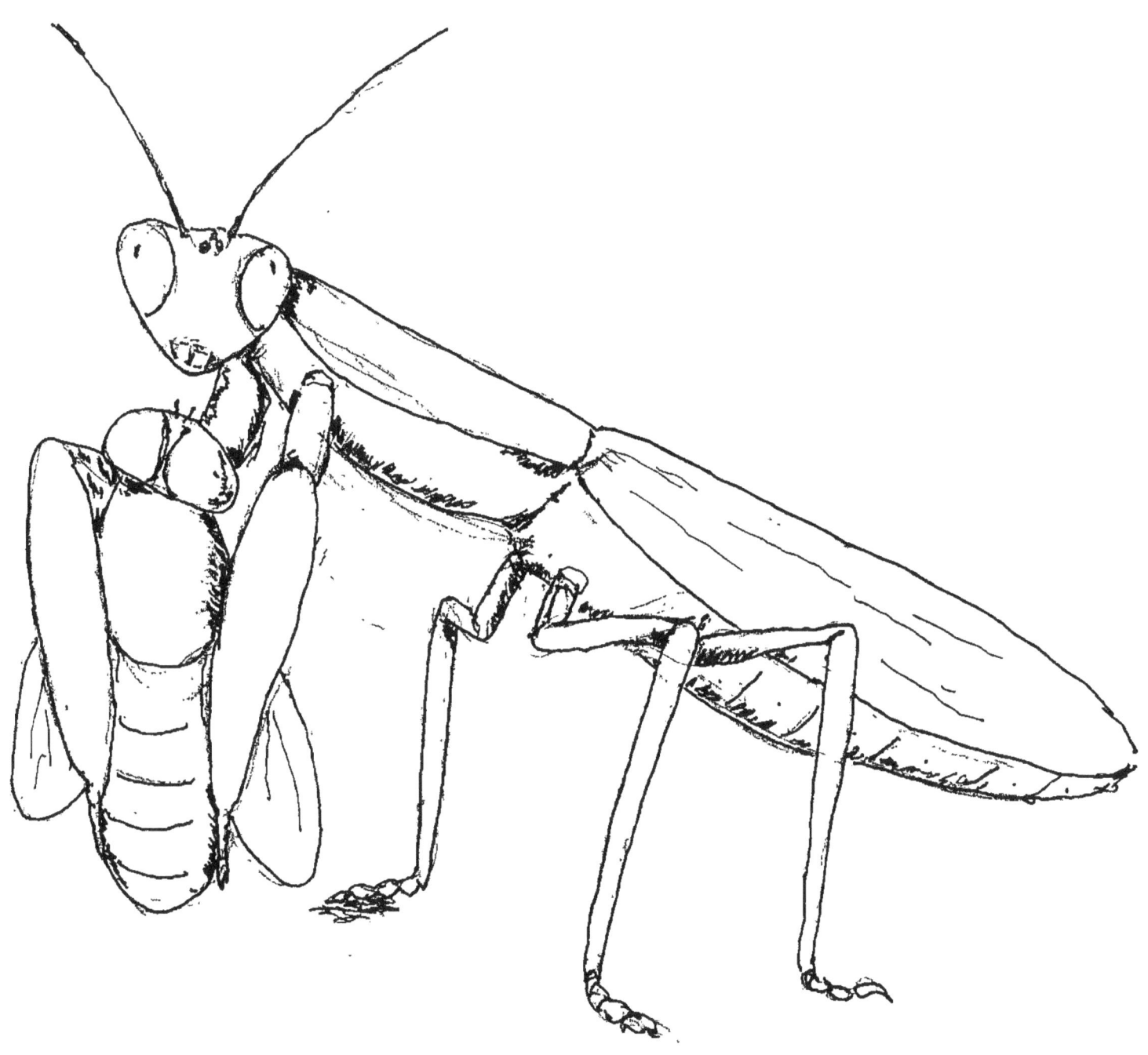

A dragonfly on a leaf.

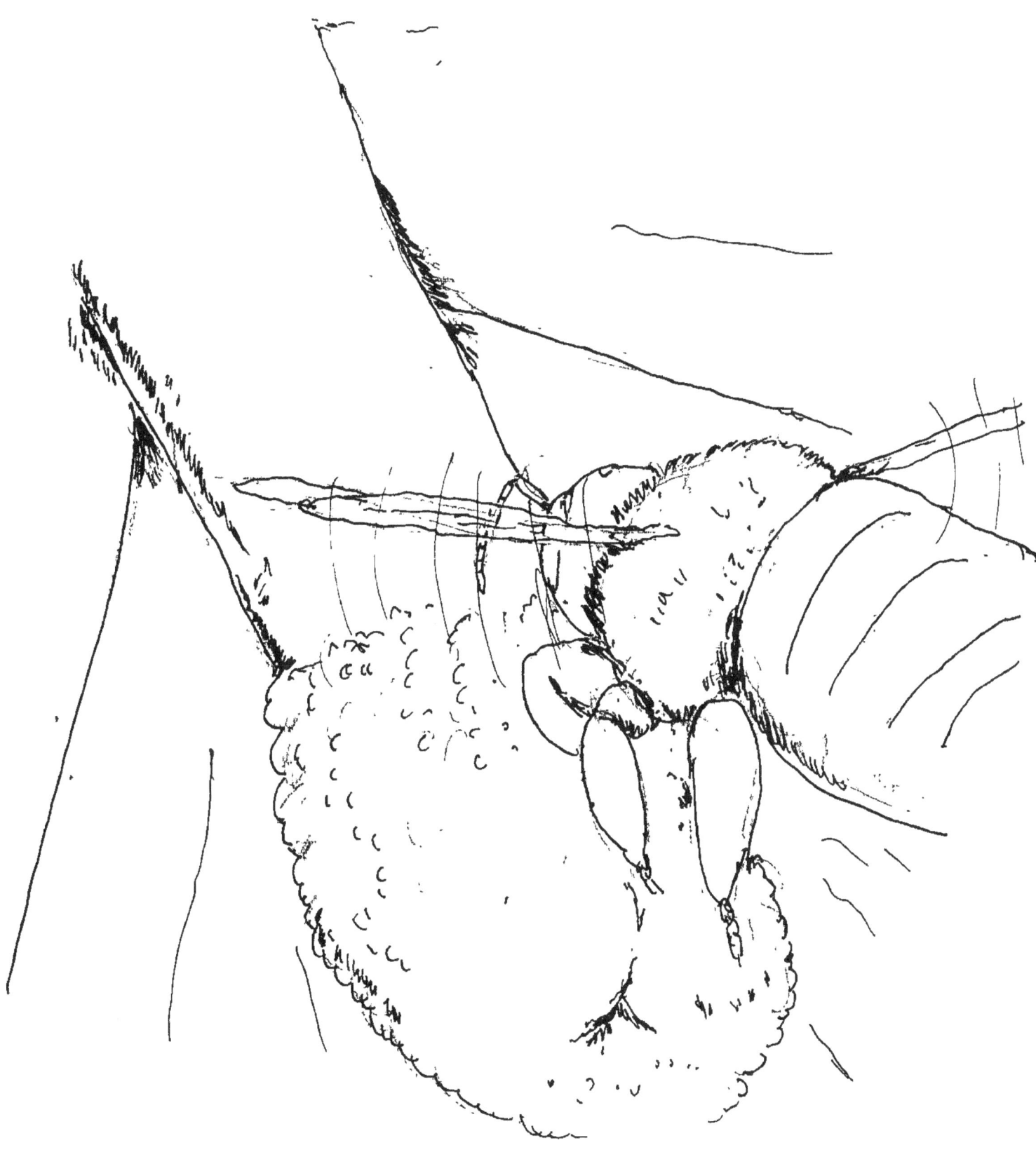

A Carpenter Bee landing on a male squash flower.

Some kinds of insects   help to control pest numbers by hunting them.

Many species of ground beetles, tiger beetles, and ladybugs (beetles) are some of the predators that help to control harmful insect pests in your garden and/or farm.

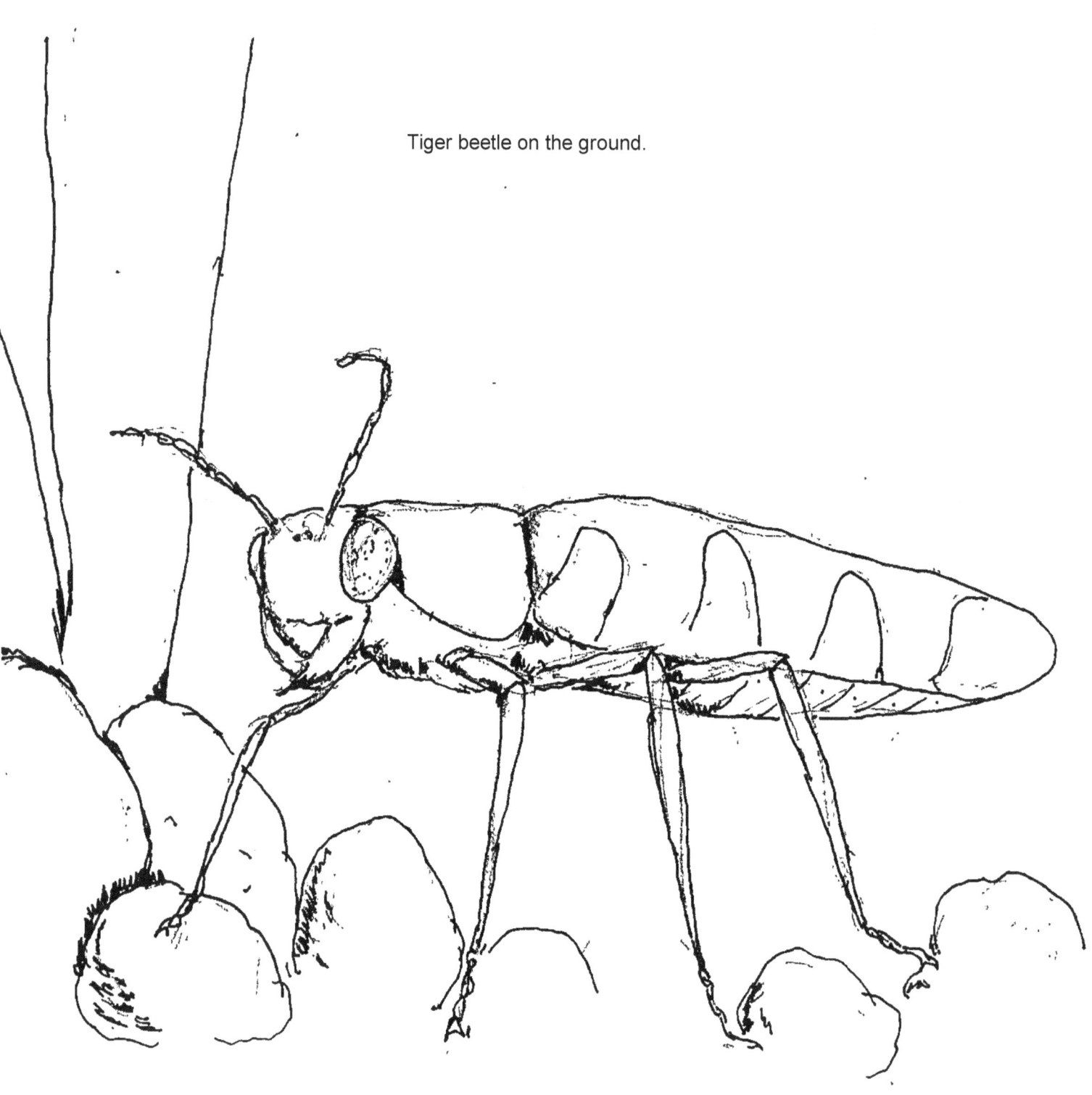

Tiger beetle on the ground.

Tiger Beetle

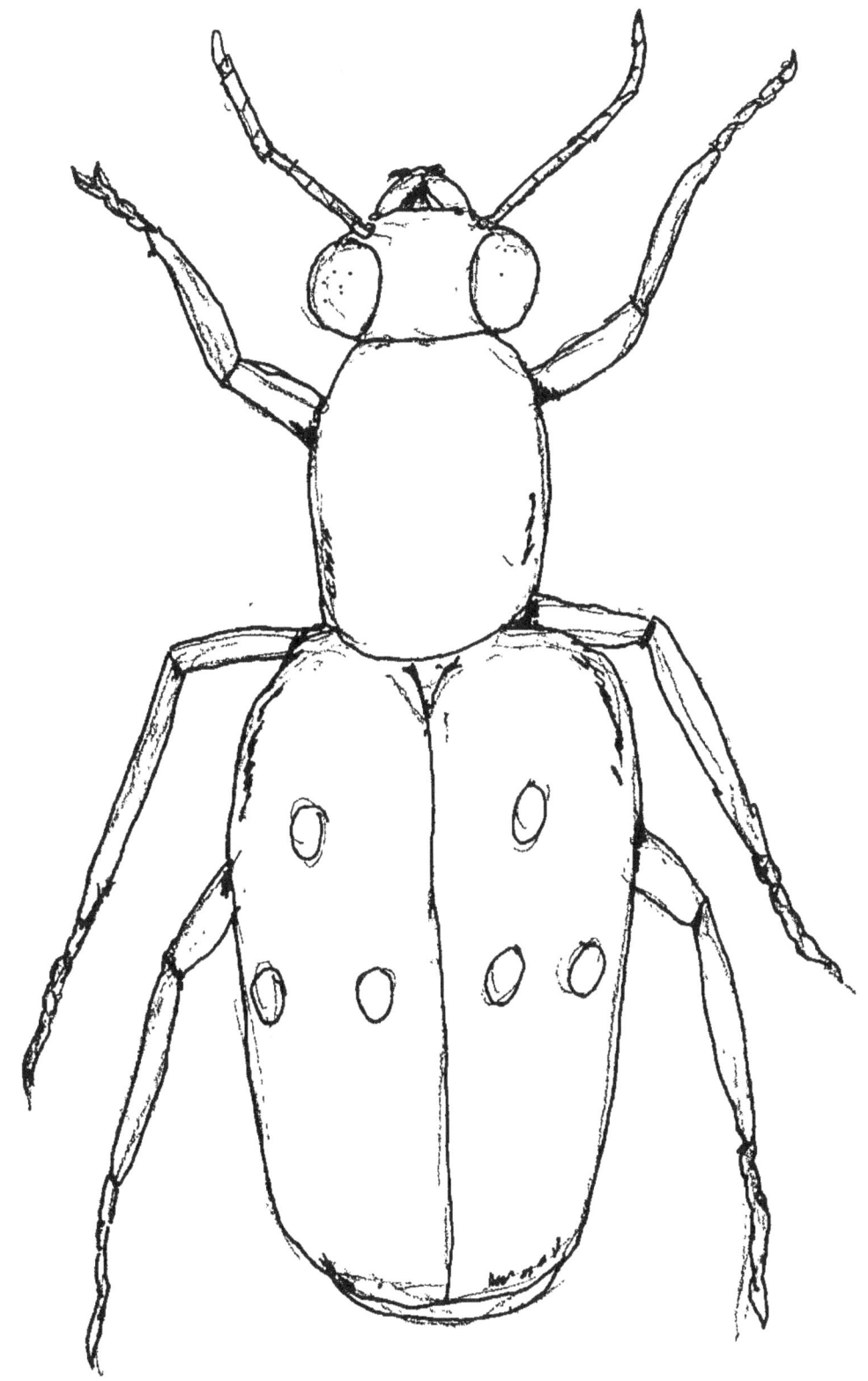

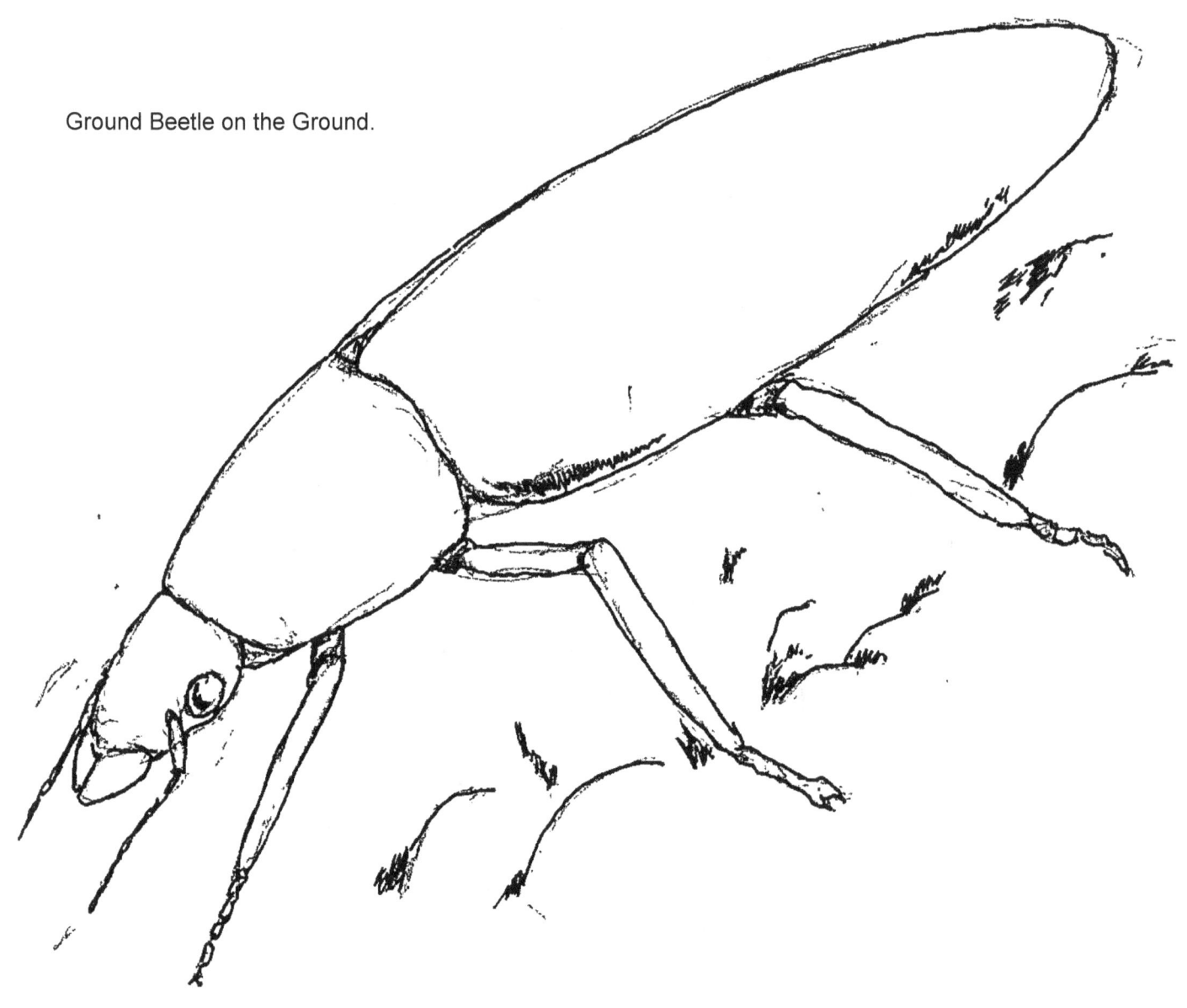

Ground Beetle on the Ground.

A Ground Beetle captured a caterpillar for lunch.

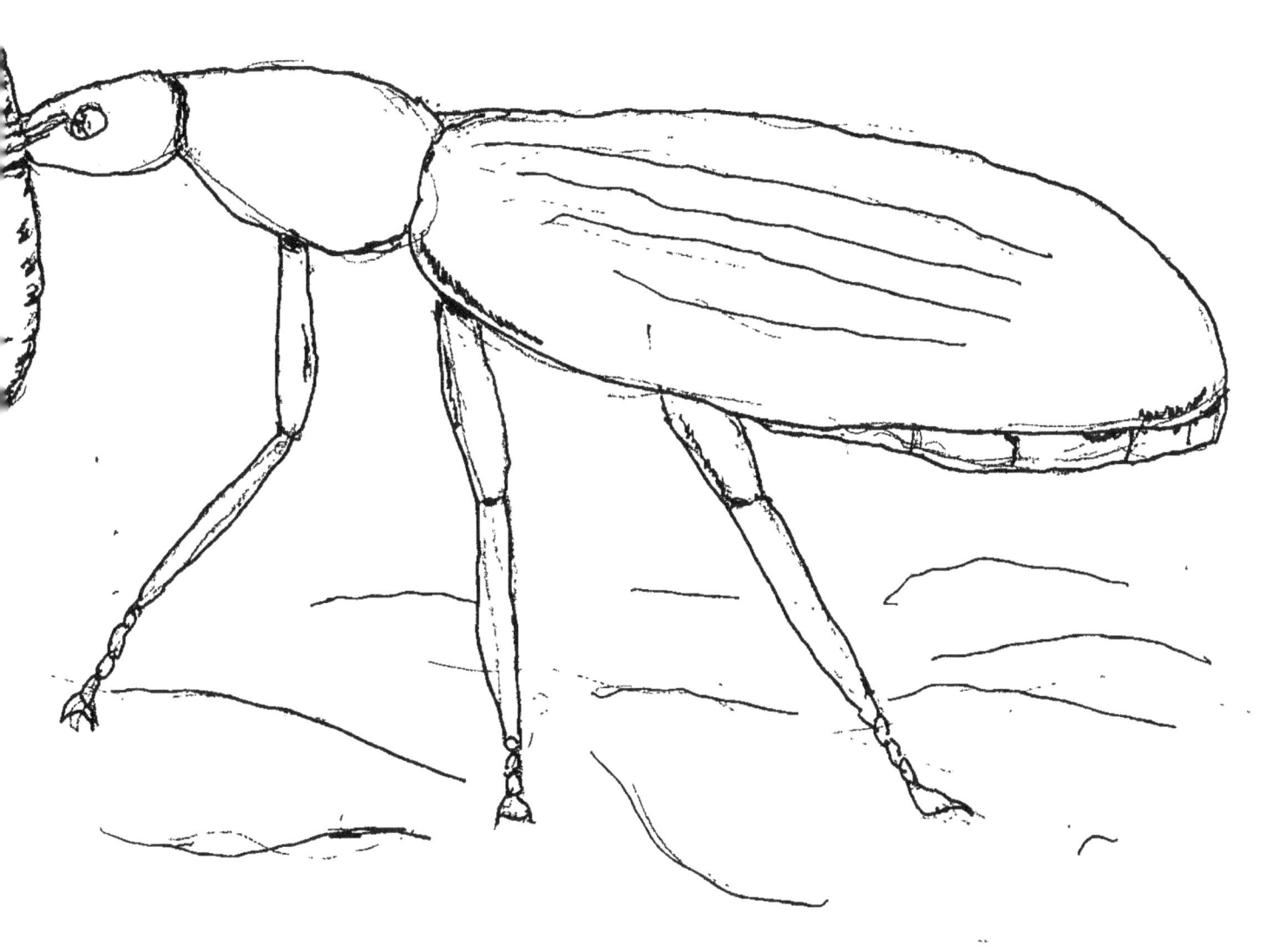

Ladybug (beetle)

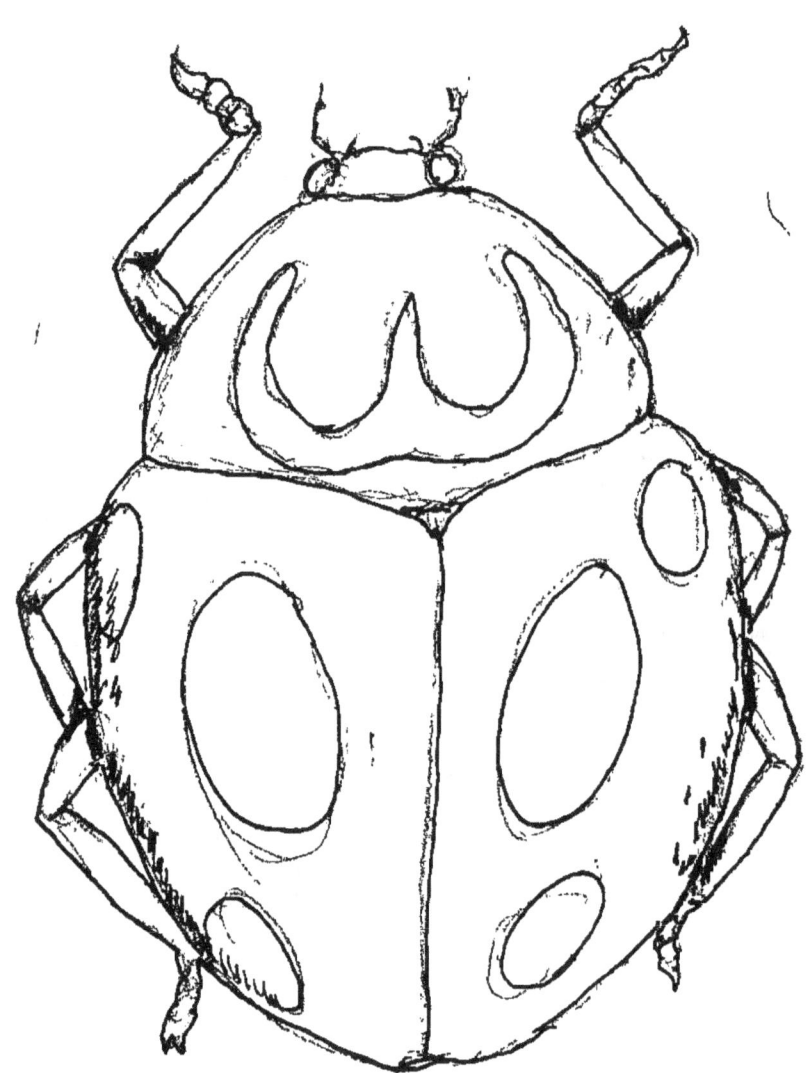

Dragonflies and damselflies feed on mosquitoes and other smaller insects. Their activities can help to control these pest populations in a certain area.

Damselfly

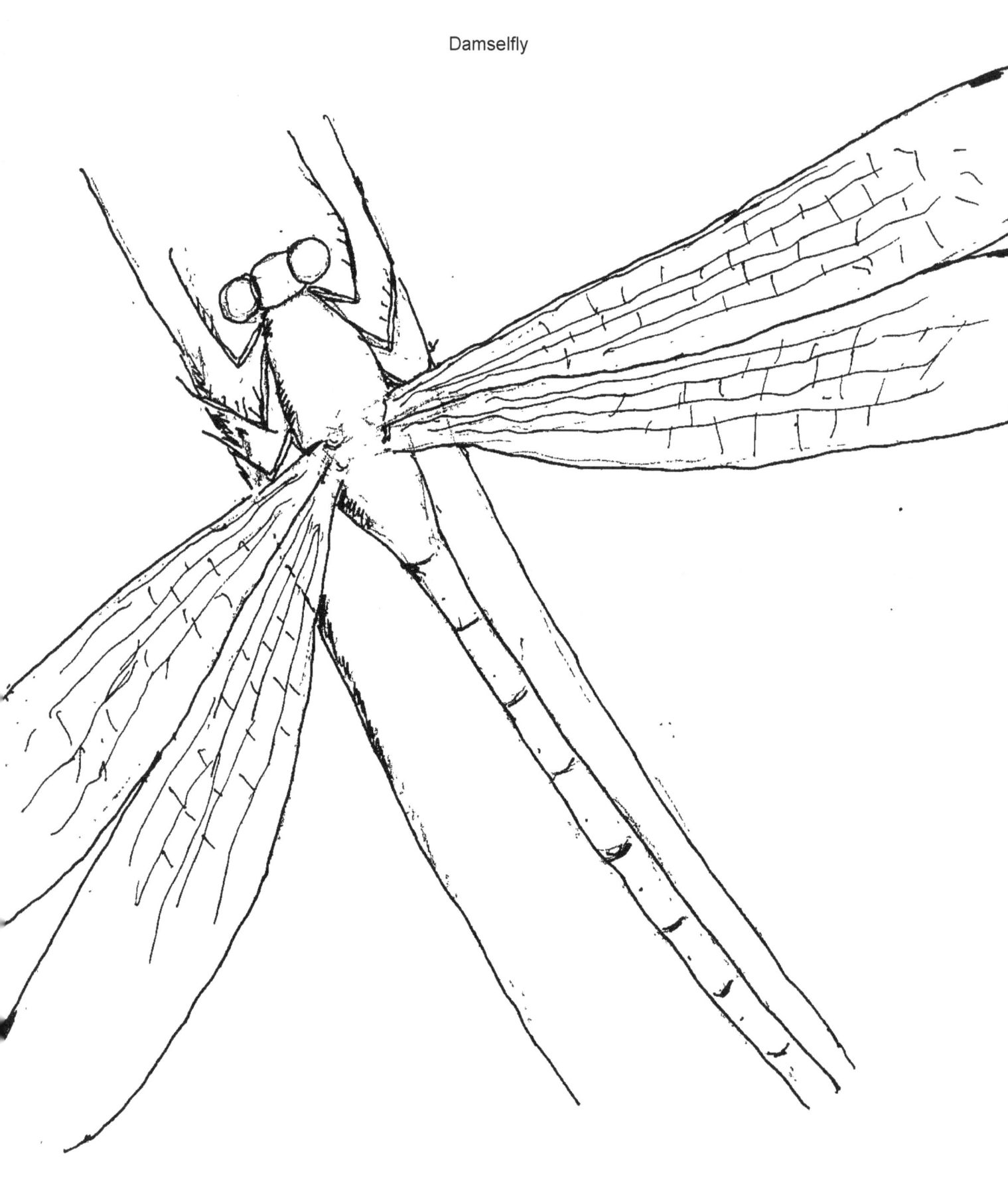

Dragonfly

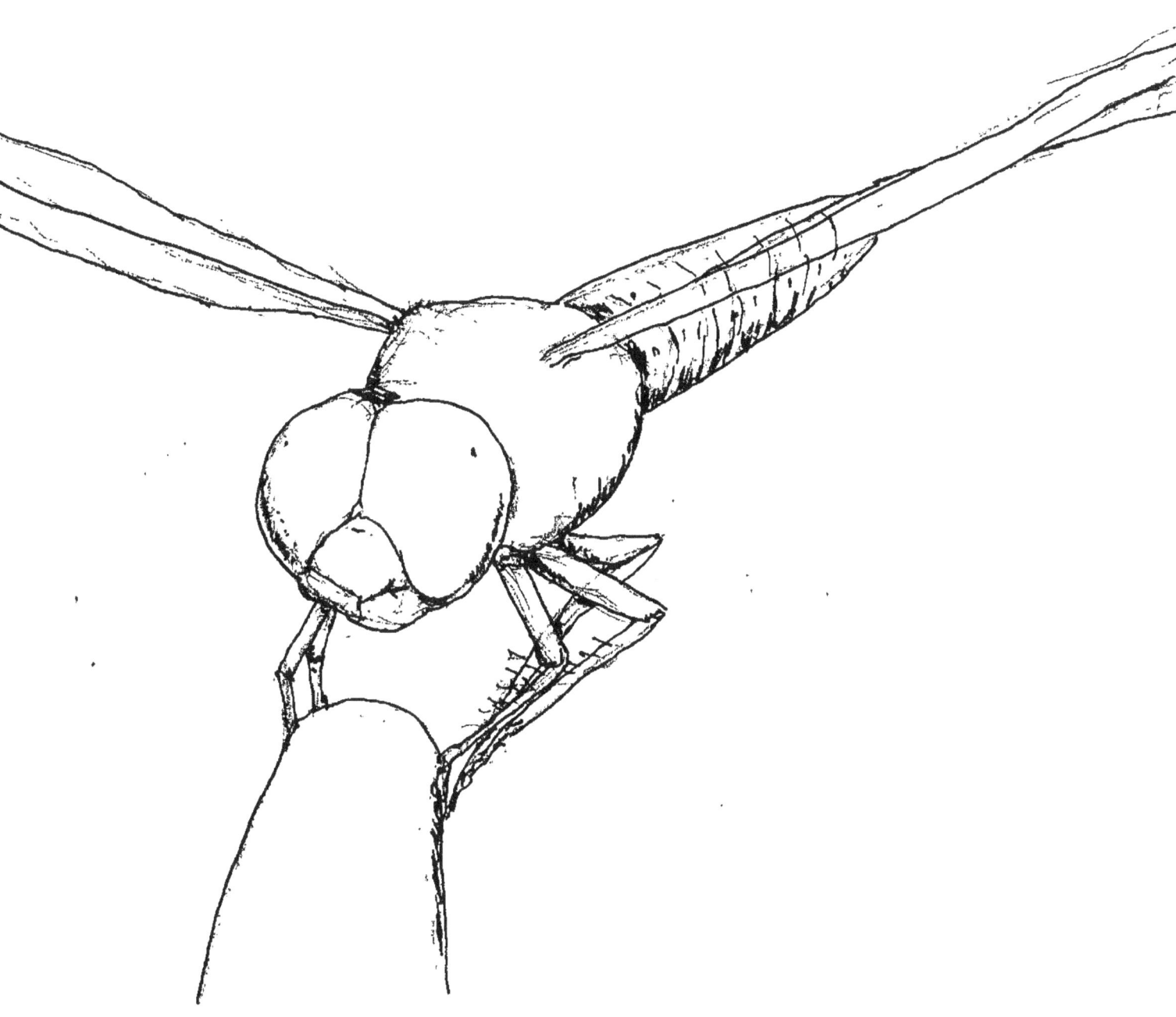

Praying mantises are another group of predators that feeds on other insects and smaller animals.

Praying mantis

A tropical mantis that is made as a leaf.

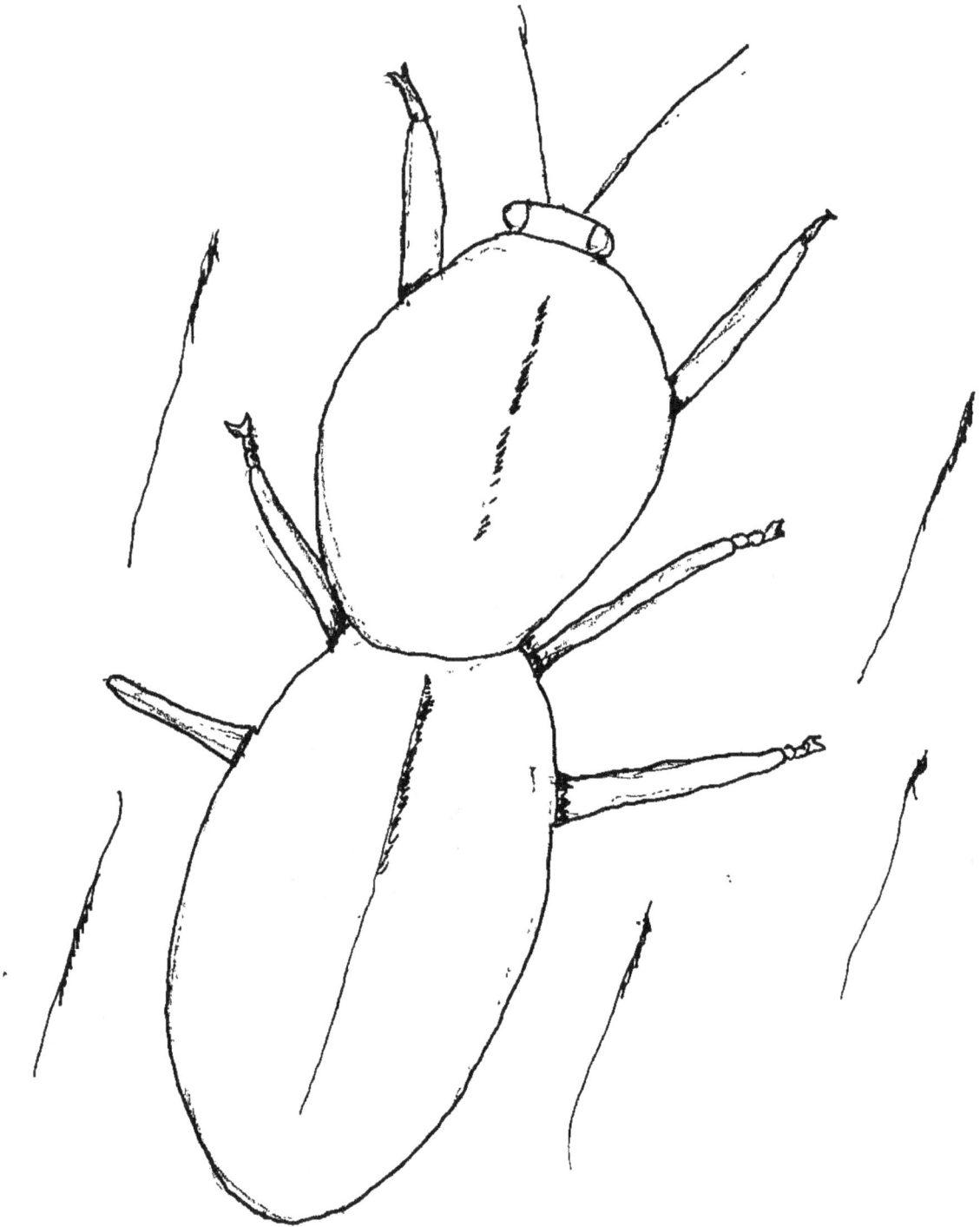

A female mantis laying an egg case.

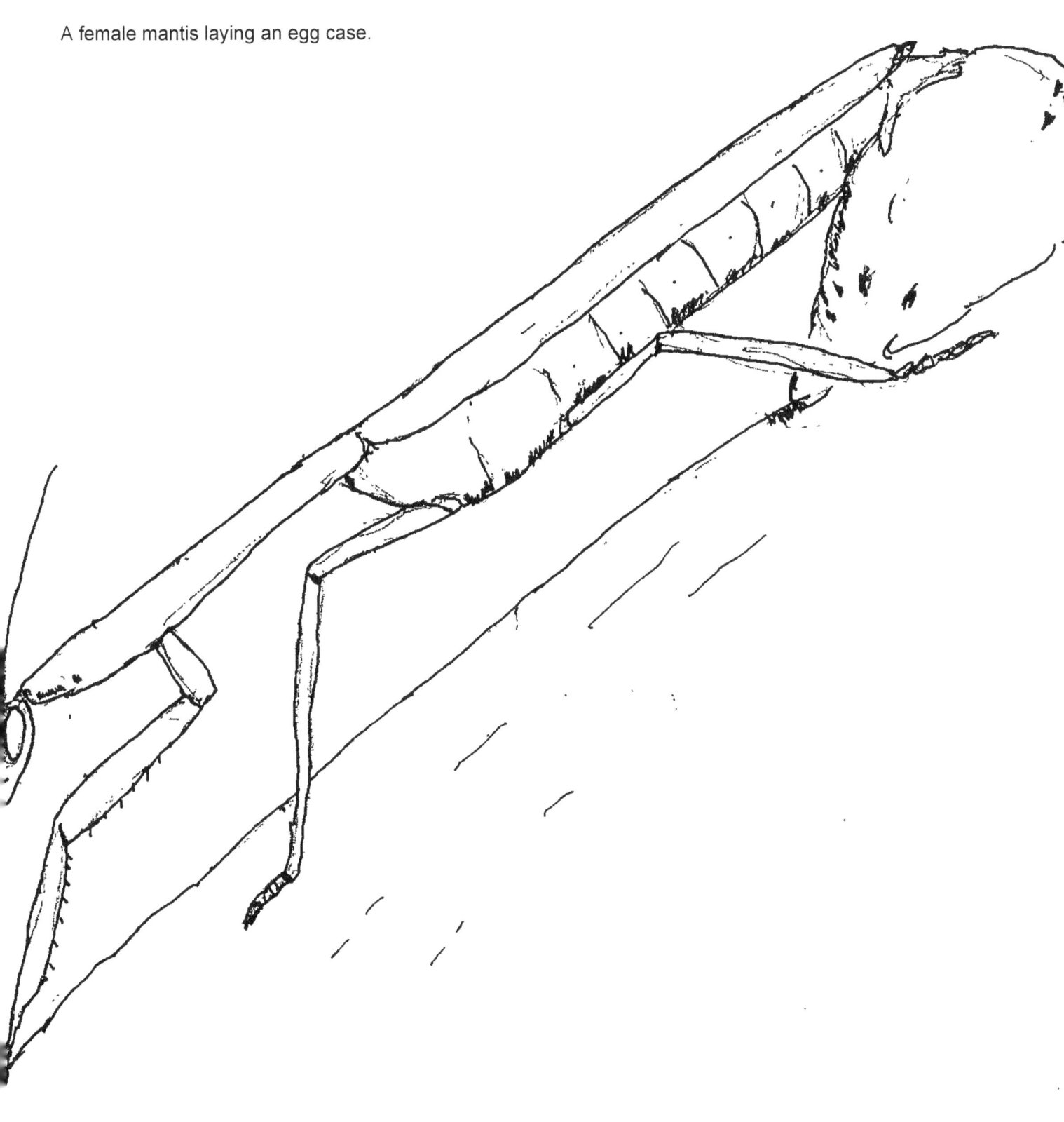

Butterflies, bees, and hoverflies are some of the insects that help to pollinate many of our crop plants by transferring pollen between flowers. They help to produce many of our favorite fruits and vegetables.

Bees and butterflies on a flower head.

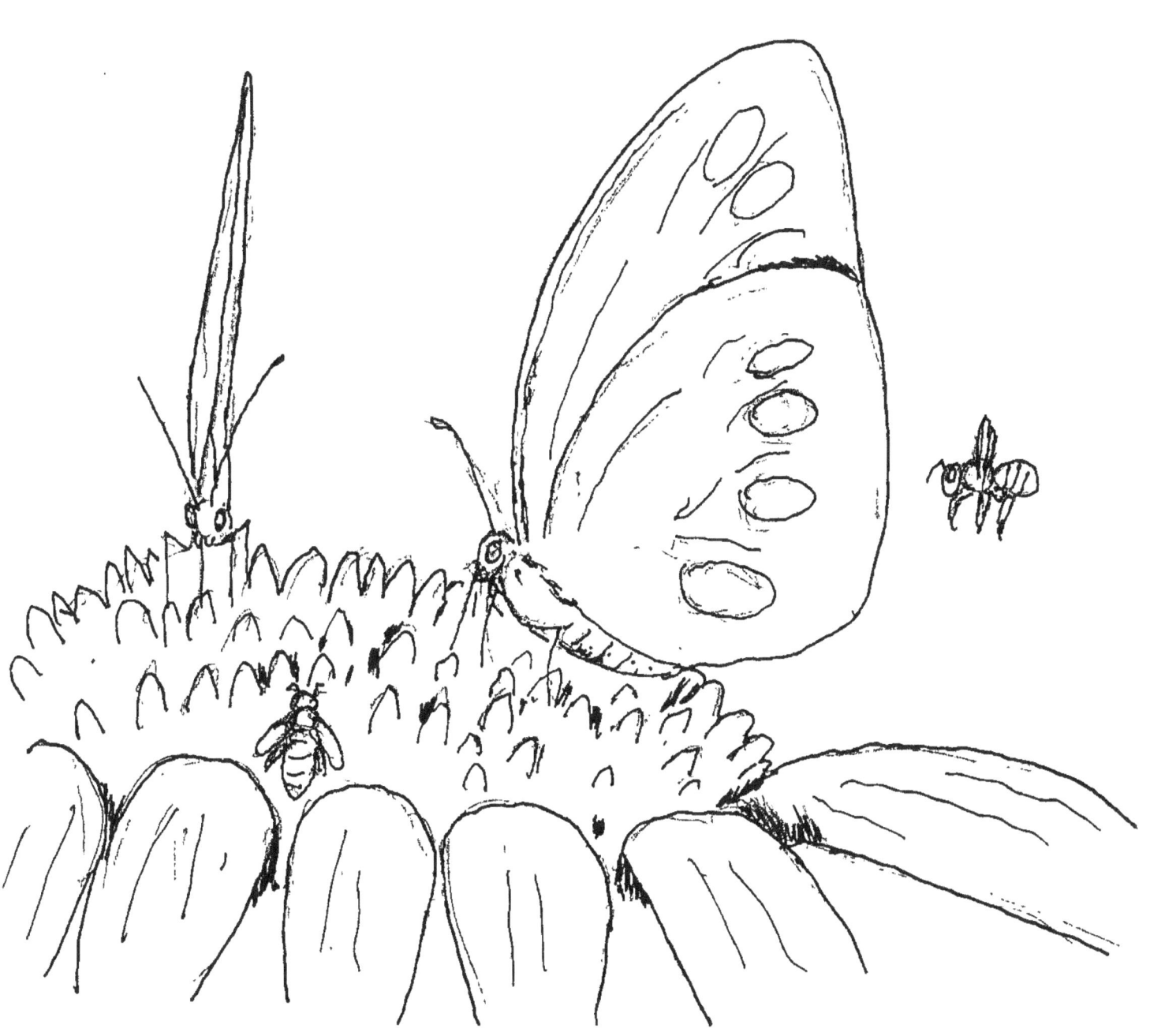

A bee landing on a flower.

A hoverfly on a strawberry flower.

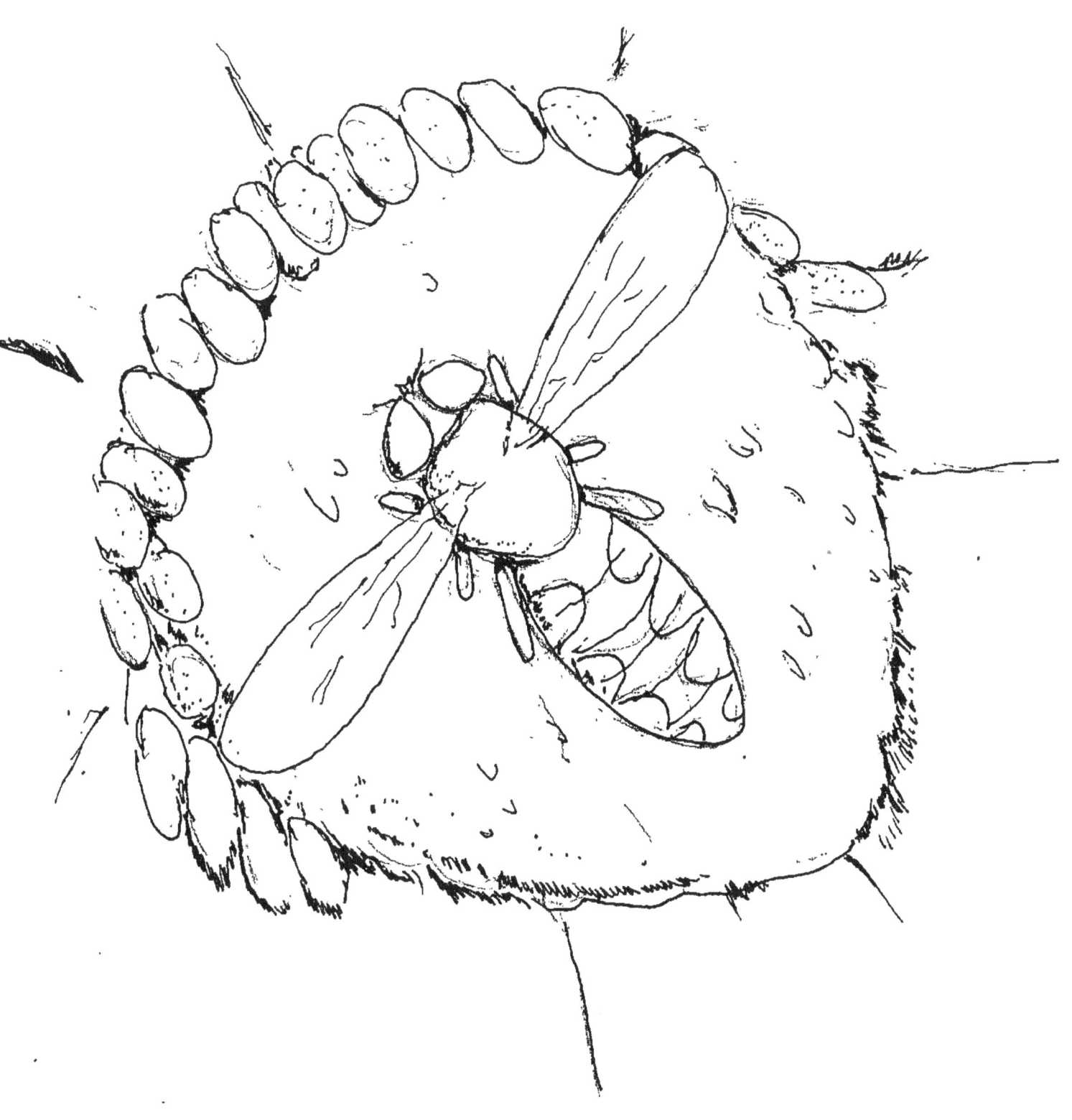

Bee

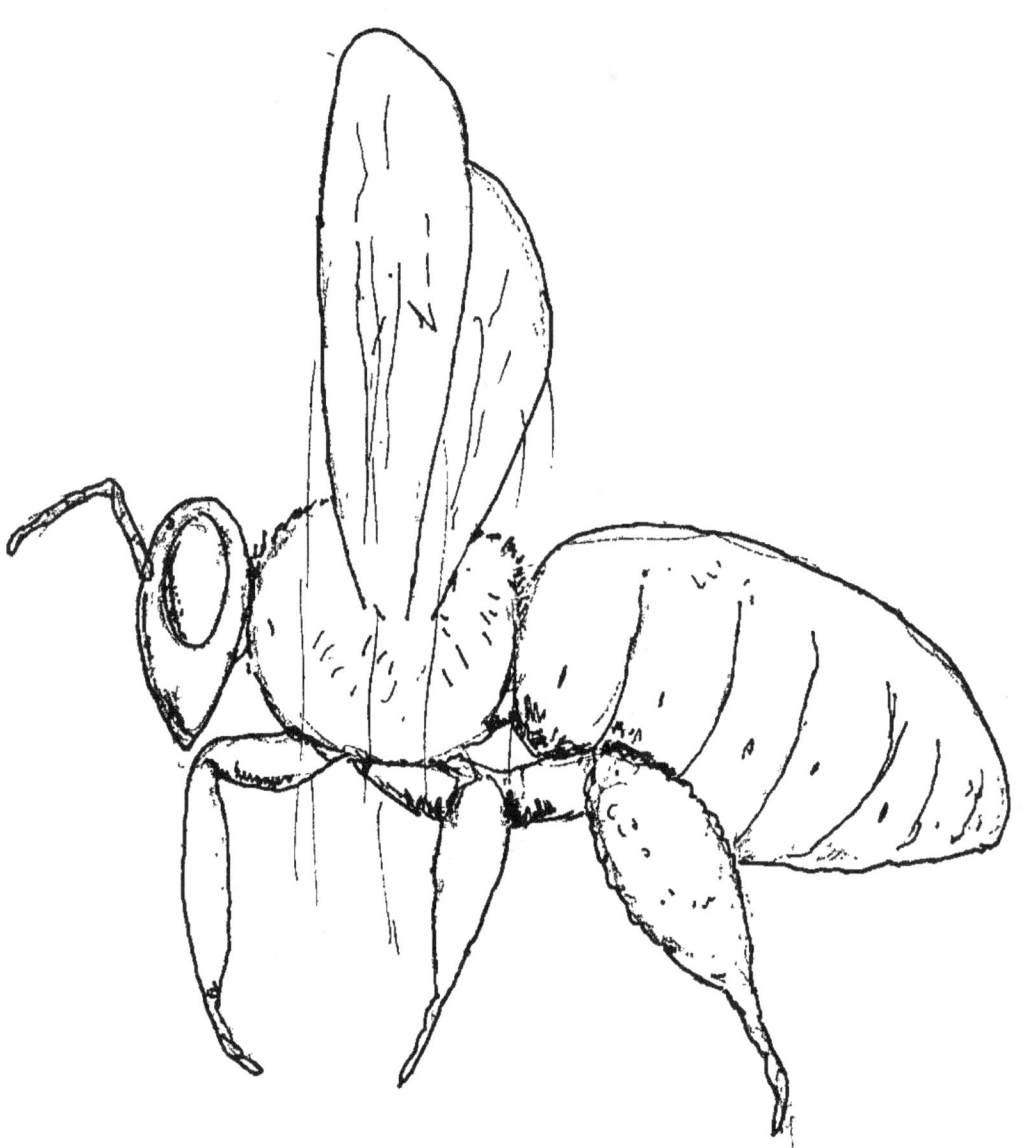

Blueberries

Tomatoes

Besides producing many of our fruits and vegetables, honey bees help to make honey and beeswax for our use.

Strawberries

Apples

Jar of honey

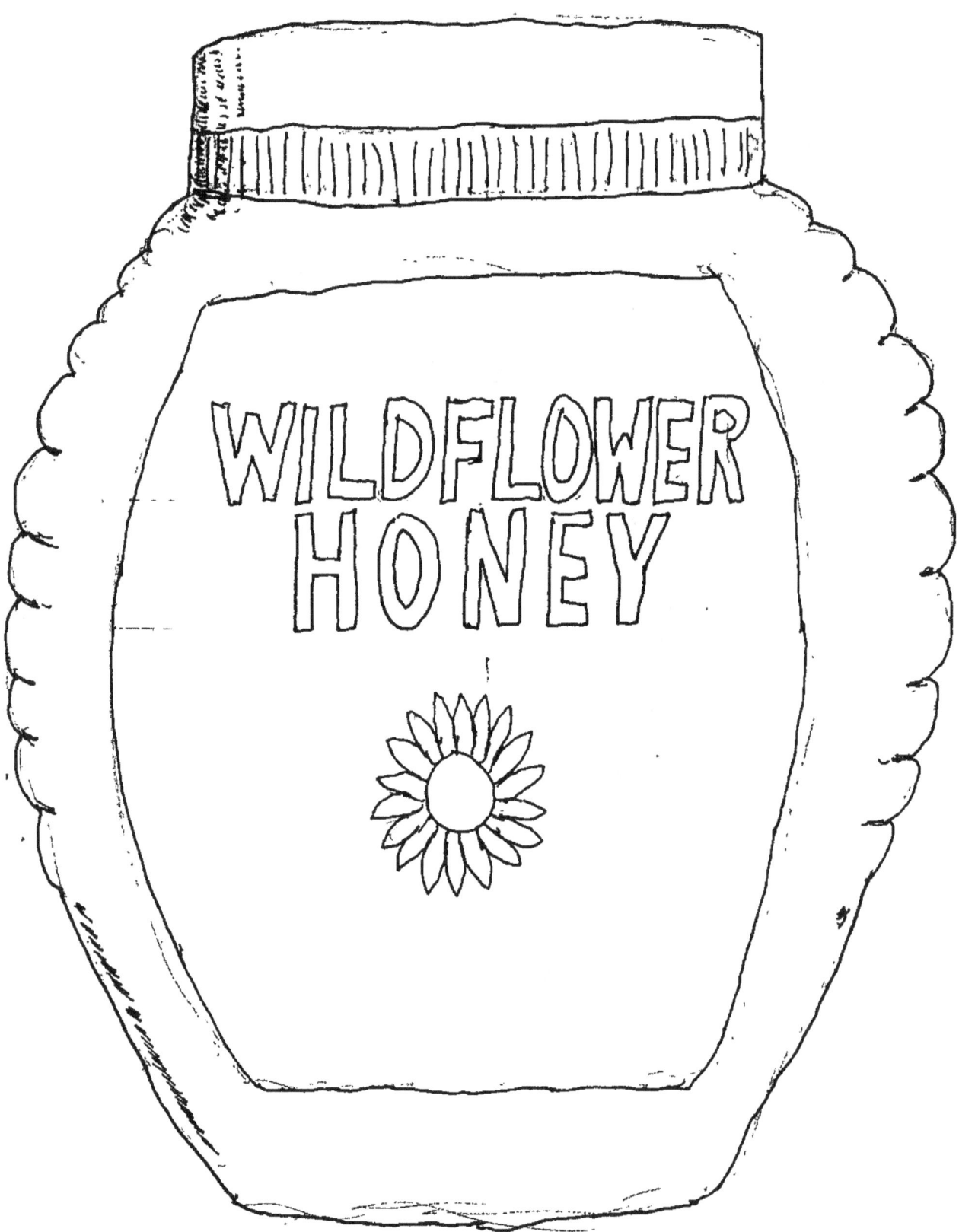

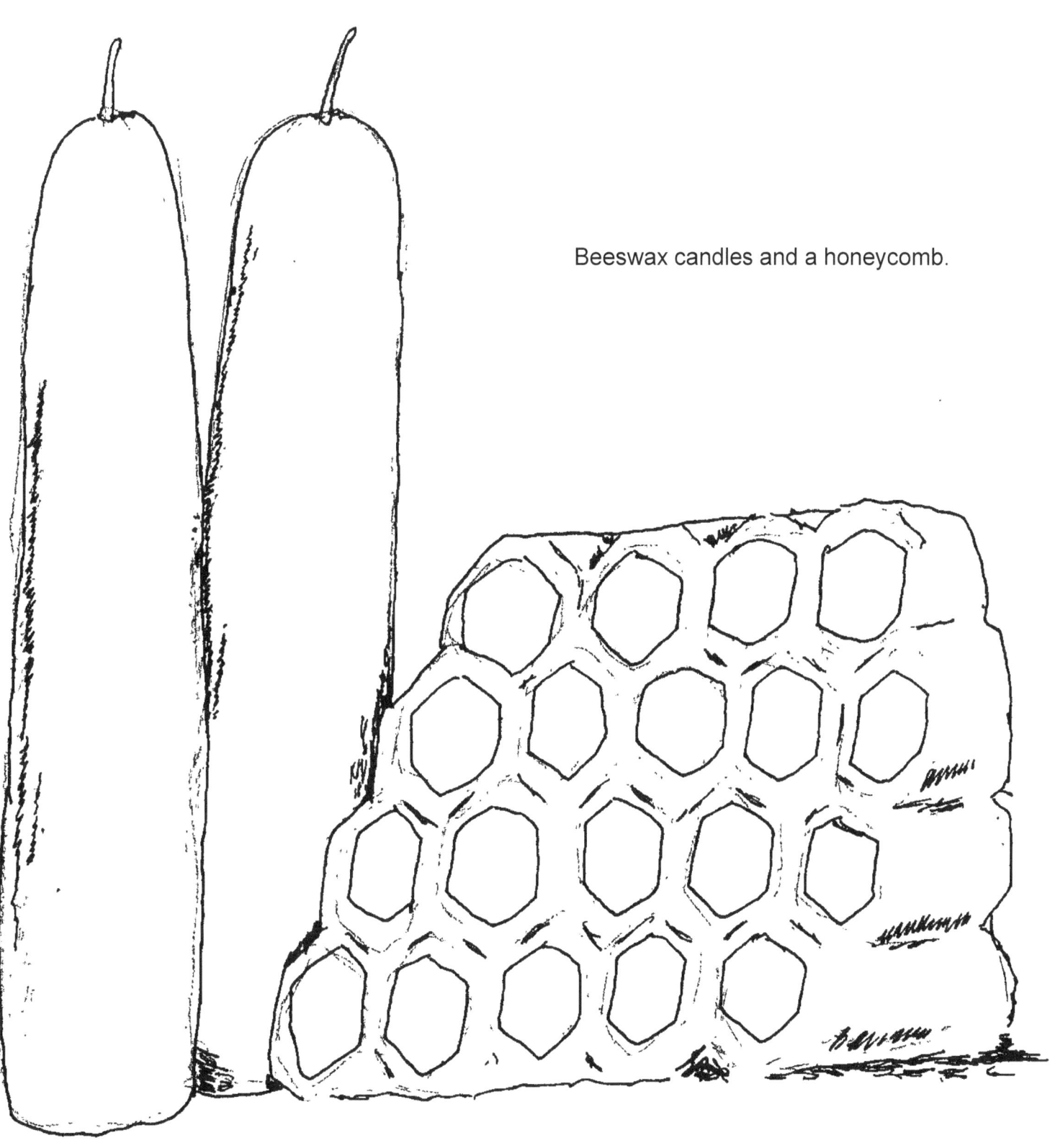

Beeswax candles and a honeycomb.

Honeybee

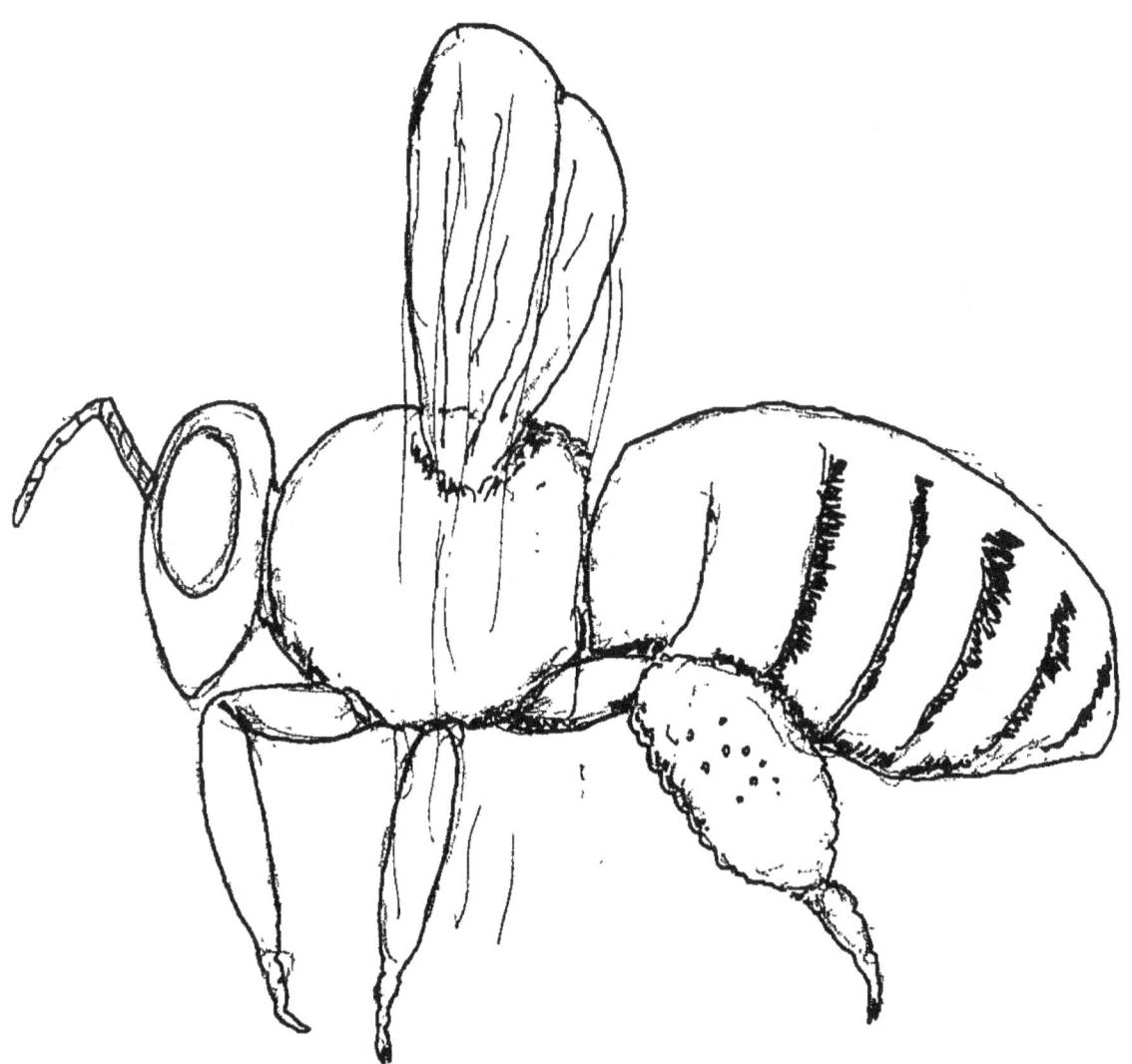

We can learn more about these and other insects through books, computers, science museums, and teachers.  However, we can turn to God for knowledge and understanding of them.

# References for Use

Imes, Rick. **The Practical Entomologist**.  Simon & Schuster, Inc. New York. 1992.

Mader, Eric et al. **Attracting Native Pollinators**.  Storey Publishing, North Adams. 2011.

Starcher, Allison Mia. **Good Bugs for Your Garden**.  Algonquin Books of Chapel Hill.    1995.

www.ingramcontent.com/pod-product-compliance
Lightning Source LLC
Chambersburg PA
CBHW081310180526
45170CB00007B/2645